MW01364854

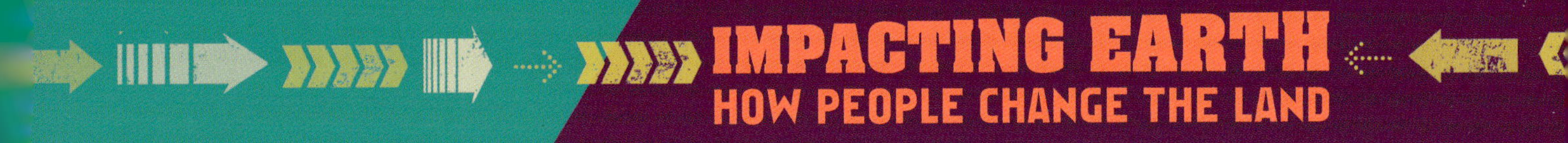

IMPACTING EARTH
HOW PEOPLE CHANGE THE LAND

BUILDING FACTORIES AND POWER PLANTS

ELIZABETH KRAJNIK

PowerKiDS press
New York

Published in 2019 by The Rosen Publishing Group, Inc.
29 East 21st Street, New York, NY 10010

First Edition

Editor: Jennifer Lombardo
Book Design: Tanya Dellaccio

Photo Credits: Cover Neil Mitchell/Shutterstock.com; p. 5 airphoto.gr/Shutterstock.com; p. 7 MPI/Archive Photos/Getty Images; p. 9 Igor Grochev/Shutterstock.com; p. 11 Peter Turnley/Corbis Historical/Getty Images; p. 13 altrendo images/Stockbyte/Getty Images; p. 15 euregiocontent/Shutterstock.com; p. 16 small smiles/Shutterstock.com; p. 17 kamilpetran/Shutterstock.com; p. 19 Migel/Shutterstock.com; p. 21(main) fokke baarssen/Shutterstock.com; p. 21 (inset) SkyLynx/Shutterstock.com; p. 22 Gary Whitton/Shutterstock.com.

Library of Congress Cataloging-in-Publication Data

Names: Krajnik, Elizabeth, author.
Title: Building factories and power plants / Elizabeth Krajnik.
Description: New York : PowerKids Press, [2019] | Series: Impacting earth : how people change the land | Includes bibliographical references and index.
Identifiers: LCCN 2018023977| ISBN 9781538341834 (library bound) | ISBN 9781538341810 (pbk.) | ISBN 9781538341827 (6 pack)
Subjects: LCSH: Factories–Juvenile literature. | Power-plants–Juvenile literature.
Classification: LCC TH4511 .K68 2019 | DDC 690/.54–dc23
LC record available at https://lccn.loc.gov/2018023977

CPSIA Compliance Information: Batch #CWPK19. For Further Information contact Rosen Publishing, New York, New York at 1-800-237-9932

CONTENTS

WHY WE NEED FACTORIES AND POWER PLANTS

Many years ago, life on Earth was much different. People had to use candles to see in the dark. They stayed warm by burning wood or coal in a fireplace, which made the air in people's houses hard to breathe. Today, most people use **electricity** to light and sometimes heat their homes. Factories and power plants help them do this.

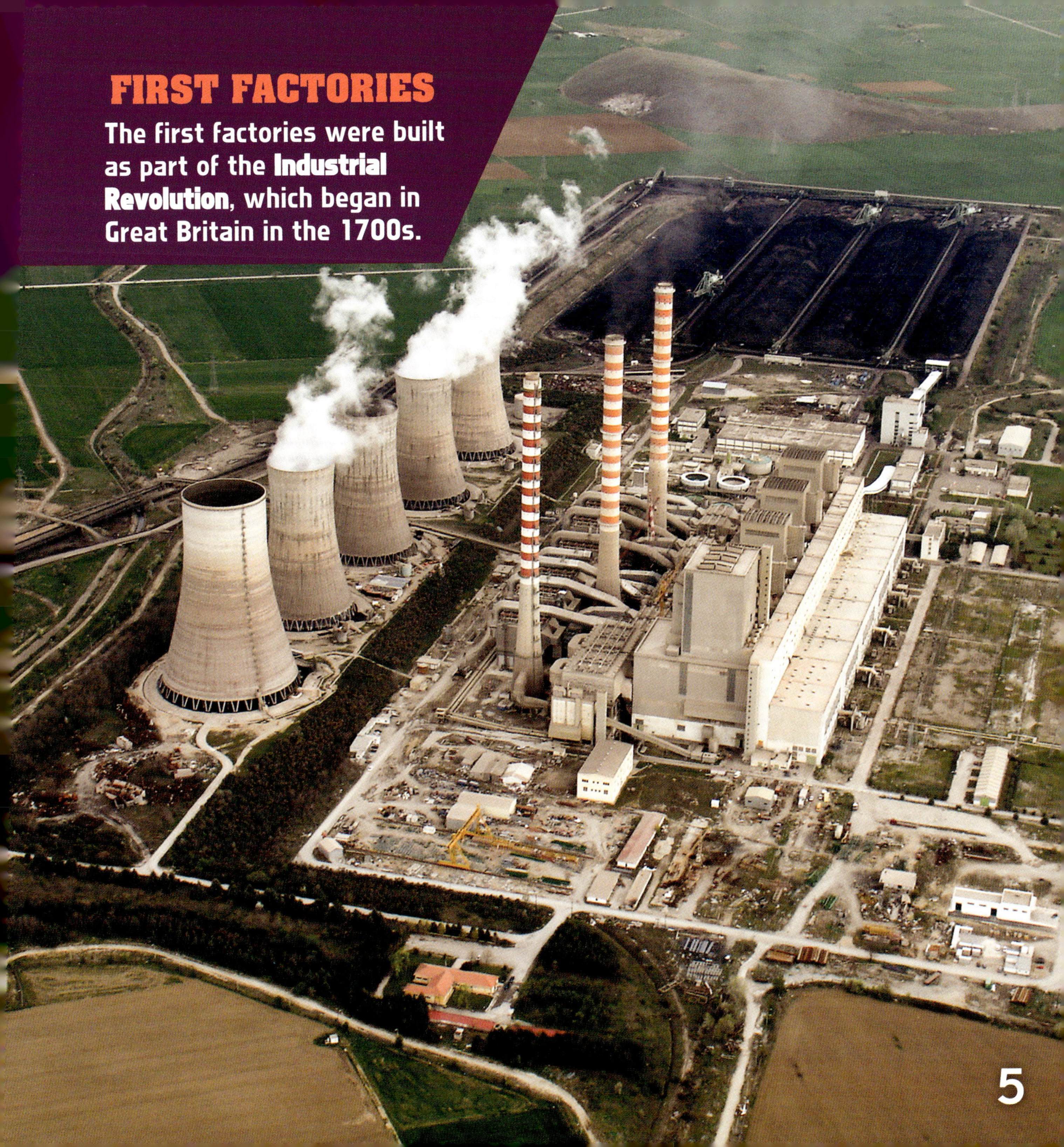

FIRST FACTORIES

The first factories were built as part of the **Industrial Revolution**, which began in Great Britain in the 1700s.

THE WORLD POPULATION RISES

Long ago, most people depended on farming to live. However, as the number of people in the world rose, many moved to cities to find work. Better farm machinery also allowed farmers to make the land produce more food to support the people living in cities. However, these new ways of life weren't always better. People began to make huge changes to Earth's land as farms and cities grew larger.

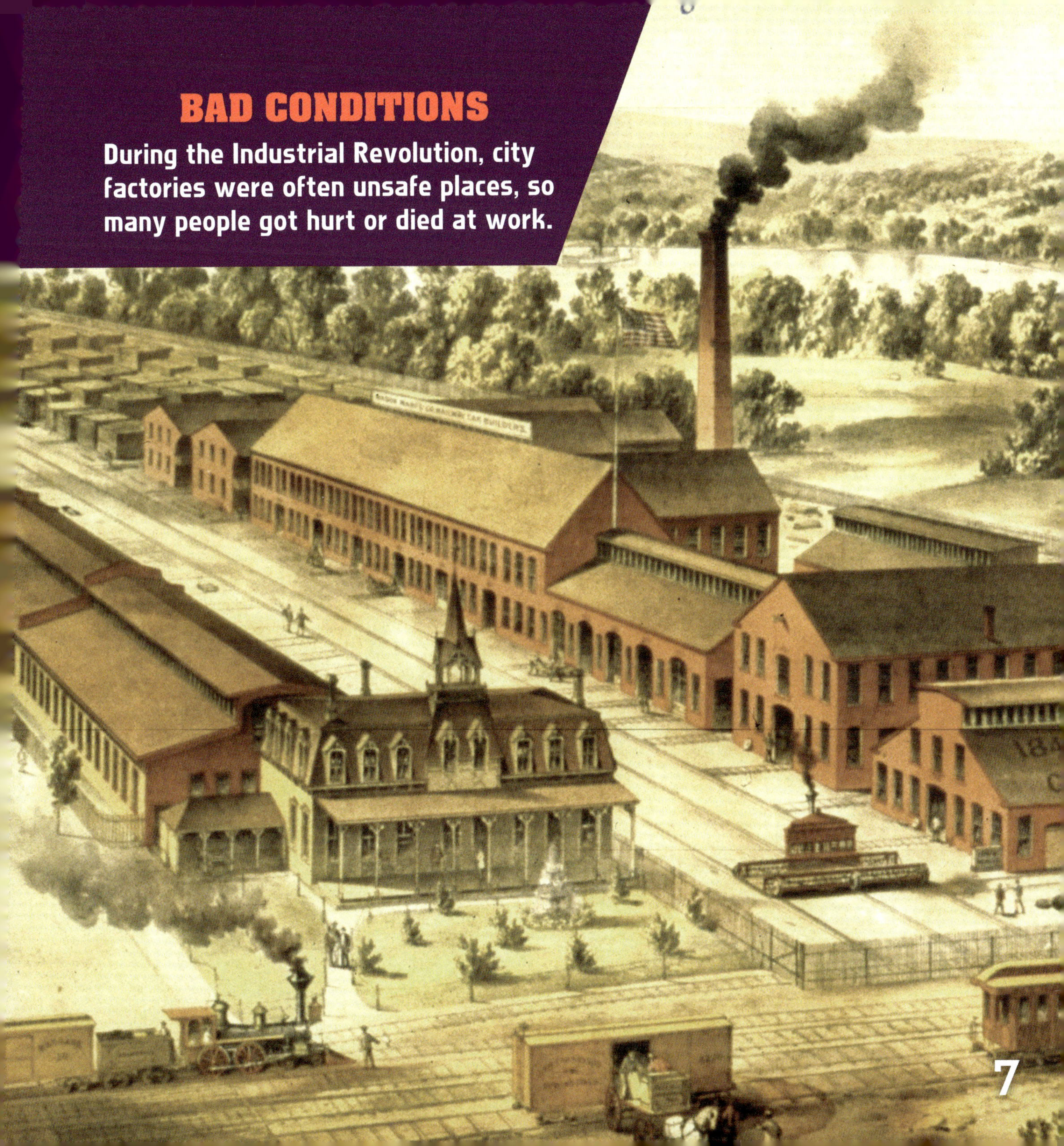

BAD CONDITIONS

During the Industrial Revolution, city factories were often unsafe places, so many people got hurt or died at work.

TYPES OF FACTORIES

Today, there are many different types of factories. There are factories that make car parts, clothes, paper, or computers. Each of these factories needs different things to make its products. Paper mills need wood pulp, or soft, soggy bits of wood. They also need lots of water and **energy** to make paper. Clothing factories need energy to power weaving machines called looms, which turn yarn or thread into cloth.

PAPER MILL

HARMING THE EARTH

Factories use **resources** such as trees from the land, but they often also add things to the land that don't belong there. During the papermaking process, **chemicals** are added to wood chips to create wood pulp. These chemicals, which may be harmful, come out in the water that's used to make paper. This can make water sources dirty and unsafe.

THE NEED FOR ELECTRICITY

All factories need electricity. Without it, machines in factories wouldn't run, the lights wouldn't work, and there wouldn't be any heat. We also need electricity to power our homes and other businesses. We get electricity from power plants. There are different types of power plants, including nuclear, coal, solar, and wind power plants. Some types of power plants are better for Earth than others.

SOLAR-POWERED HOMES

Some people put solar panels on their roofs to power their homes. This can help them use fewer **fossil fuels**!

NUCLEAR POWER PLANTS

Nuclear power plants use uranium—**radioactive** matter found in the earth—to create energy. Uranium must be mined from the earth. To make a mine, people have to dig into the ground. This changes the way the land looks and harms the earth in a number of ways. Uranium can also be bad for the health of people and animals.

COAL POWER PLANTS

Coal power plants burn coal to create electricity. Coal, like uranium, must be mined. Open-pit coal mines clear the land of trees and wildlife. They also cause soil to wear away. Water carries harmful by-products of coal mining to larger bodies of water such as rivers and lakes. These by-products often make water sources dirty.

COAL

BURNING COAL

Burning coal adds harmful gases to the air around Earth. Many scientists think this causes Earth's temperatures to rise.

SOLAR POWER PLANTS

Solar power plants change light from the sun into electricity. Most do this using special cells that cover a flat surface called a solar panel. When light hits the cells, they turn it into electricity. Solar energy is renewable, meaning it will naturally be made again as it's used. Solar power plants take up a lot of space. They must be built in wide-open areas where it's almost always sunny.

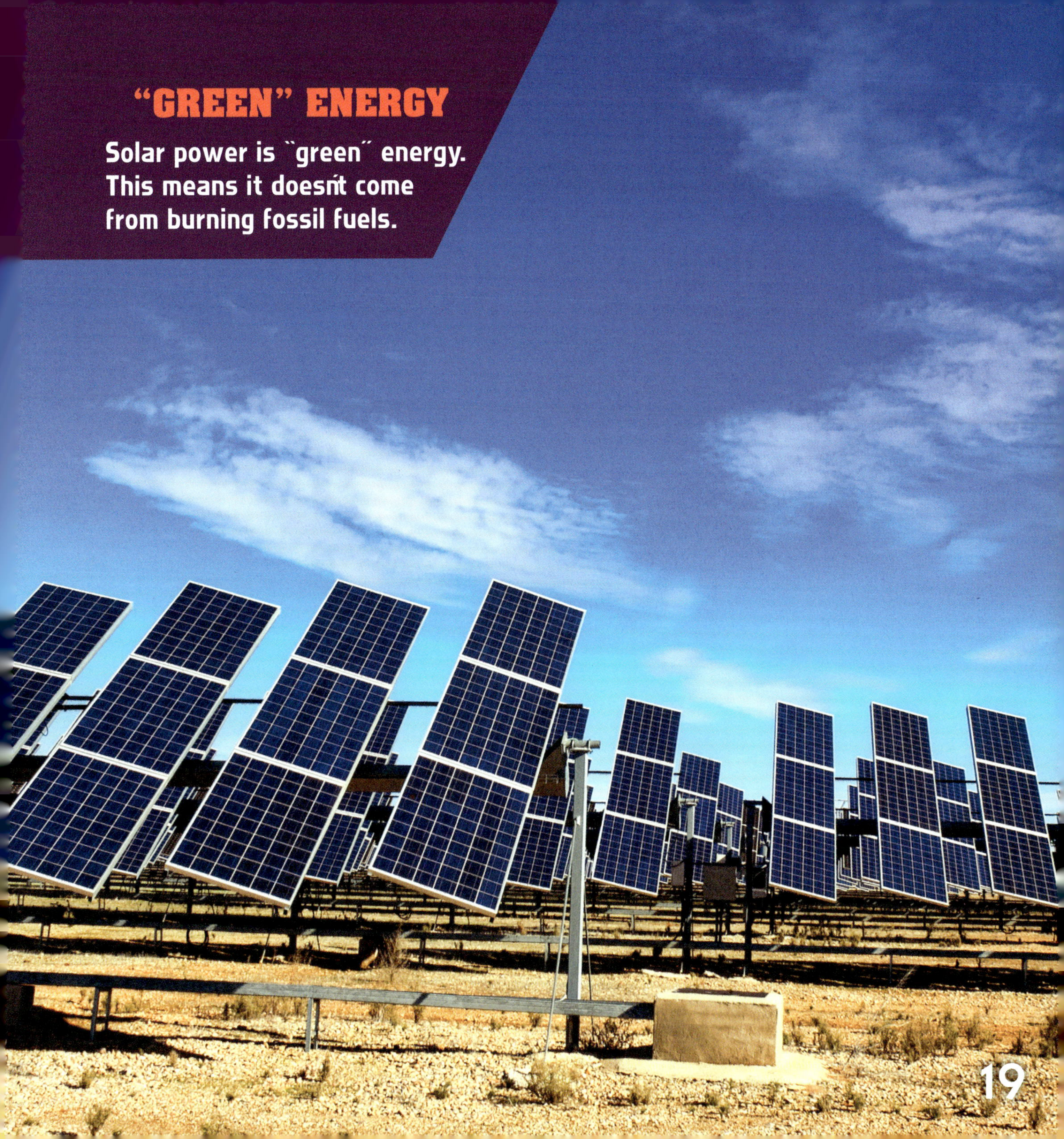

"GREEN" ENERGY

Solar power is "green" energy. This means it doesn't come from burning fossil fuels.

WIND POWER PLANTS

Wind energy is another type of renewable resource. Wind farms are made up of wind turbines, which are like tall fans with huge spinning blades. The turbine creates electricity when the wind blows. Wind farms take up a lot of space, and some people don't like the way they make the land or water look.

GOOD AND BAD

We need factories so we have things such as clothes and paper. We need power plants to power our homes and businesses. However, with the good also comes the bad. Factories and power plants often create harmful by-products and take many natural resources from the earth. We need to work hard to keep our planet healthy.

GLOSSARY

chemical: Matter that can be mixed with other matter to cause changes.

electricity: The flow of energy made up of small particles called electrons.

energy: The power to work or to act.

fossil fuel: A fuel—such as coal, oil, or natural gas—that is formed in the earth over millions of years from dead plants or animals.

Industrial Revolution: An era of social and economic changes marked by advances in technology and science.

radioactive: Having or producing a powerful and dangerous form of energy called radiation.

resource: Something that can be used.

INDEX

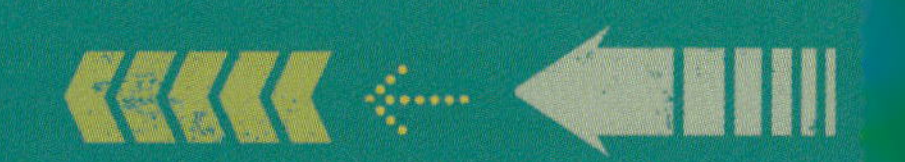

WEBSITES

Due to the changing nature of Internet links, PowerKids Press has developed an online list of websites related to the subject of this book. This site is updated regularly. Please use this link to access the list:
www.powerkidslinks.com/hpcl/factories